Dearest Mother
with lots
of love x

Lucinda

Portraits of
Astronomers

Lucinda Douglas-Menzies

Foreword
Lucinda Douglas-Menzies

Introduction
Alison Boyle

Afterword
Anthony de Jasay

The Science Museum, London in association
with Lucinda Douglas-Menzies

First published in Great Britain 2009
by Lucinda Douglas-Menzies, London

Photographs copyright © Lucinda Douglas-Menzies
Text copyright © of individual named contributors

ISBN 978-0-9561390-0-9

Designed by Derek Westwood

Printed by Cambridge University Press

For Tony, George
and Theodora

Contents

Acknowledgements

First of all I would like to thank each of the astronomers for allowing me to photograph them and for their written accounts to accompany their portraits. Without their generous collaboration this project would never have got off the ground.

Many people gave their help and advice and I would specially like to thank the following who greatly assisted me: Keith Moore at the Royal Society, David Elliot at the RAS, Sir Patrick Moore, Prof Mark Bailey, Dr Margaret Penston, Lord Martin Rees, Dame Jocelyn Bell Burnell, Prof John Barrow, Prof Ian Robson, and Pedro Russo. Special thanks goes to Alison Boyle at The Science Museum, London for her enthusiasm and assistance in the acquisition of the set of portraits and for agreeing to write the introduction to the book.

I would like to thank the following for arranging and hosting the travelling exhibition: Dr Felicity Henderson, The Royal Society, Prof Carlos Frenk , Ogden Centre, Durham University, Prof John Zarnecki,, The Open University, Greer Ramsey, Armagh County Museum, Shelley Bolderstone, Cambridge Science Festival, and Fiona Logue and Jen Wood Edinburgh Science Festival. It was a great honour to be able to have the exhibition launched from the Royal Astronomical Society in December 2008 thanks to David Elliot. I am indebted to Terence Pepper and Clare Freestone at the National Portrait Gallery, for supporting my work over the years and for arranging for portraits of ten astronomers to be on show at the NPG during the coming year. I would also like to thank James Holloway at the Scottish National Portrait Gallery for finding a home for the Scottish astronomers.

For the production of the exhibition prints I would like to thank Judy Wong at Harman Technology, Klaus Kalde, for expertly developing roll after roll of film and contact sheets, Dave Than and Mike Owen at Four Corners for the use of their magnificent darkroom, and Adrian Davidson for his superb framing. Heba Women's Project provided me the use of my own darkroom and The Wyvern Bindery created an inspired and beautiful portfolio box.

For the 'Portraits of Astronomers' book a huge thank you to Derek Westwood for his expert eye, and for the elegant design, Duncan Roeser, Russell Webb and Philip Bower at Cambridge University Press for the superb printing and Ben Backhouse at Rapid Eye for the great scans. My thanks go to Lord Martin Rees for nobly agreeing to his portrait being used on the cover, and Dr Catherine Cesarsky, for reviewing the book. I am also indebted to John Allan for his technical expertise in setting up sales of the book on my website which he designed, and Debbie Bloxam, Jeremiah Solak and Mark Bezodis from the Science Museum who all assisted me greatly with getting the book published. Chanu Miah from the SSBA gave invaluable IT assistance in times of need.

My family was a constant source of encouragement and support and I would specially like to thank my mother, Catherine Whitworth Jones for her generosity in enabling this book to be printed, my sister Arrabella Douglas-Menzies for her sound advice and my cousin Matilda Bevan for her artistic eye and help with decisions. I was lucky to have friends who offered their help and suggestions and would specially like to thank Mary Anne Kratovil, Nic Barlow, Felix Pryor, Hugh Sebag Montefiore and Lea Gratch. Brenda Stones, Jonathan and Mary Heale, and my aunt, Margaret Gascoigne, all offered me warmth and hospitality on my circuitous travels around the country. Arranging sittings with so many busy people was not an easy task and could not have happened without the help of the following: Sarah Burbidge, Judith Croadsdell, Judith Moss, Vanessa Ferraro-Wood, Tracey Ward, Dorothy Jenkins and Janet Eaton.

Finally, thank you to my husband, Tony de Jasay, for writing the afterword, and for his love and patience during the last twenty months, and my children George and Theodora for putting up with long absences from their mother.

L. D-M London 2009

Foreword
Lucinda Douglas-Menzies

In the early spring of 2007 I was looking for inspiration for a series of new portraits and happened to read a short article in *The Week* about Sir Patrick Moore. What I read about him was so remarkable that I sent off a letter asking him whether he might agree to my taking a portrait of him, which he kindly did, and that was how the portraits of astronomers began.

Patrick started me off with some names of astronomers to contact, as, coming at astronomy from the standpoint of a photographer, I knew nothing at all about the subject, apart from recognizing the Plough, and over a period of just over a year I took the thirty-eight astronomers in this book. What struck me immediately about this group of sitters was how genial and positive they were and how they loved their work so wholeheartedly. It was this refreshingly positive attitude that has driven along my own work of photographing them and made this project such a pleasure to do.

I soon realized that quite by a lucky coincidence 2009 had been declared the International Year of Astronomy to celebrate 400 years since Galileo's first observations through a telescope. It therefore seemed a natural development for me to try to tie in with the celebrations and make the portraits of astronomers into a traveling exhibition and accompanying book. Each sitter was then asked to write a short paragraph on what it was that inspired them to study astronomy to add depth and interest to their portrait.

Taking the portraits took me to places that I would never have seen on an exciting wild goose hunt: The turbulence testing laboratories at The Open University, Milton Keynes, the gargoyled roof of the Royal Observatory, Blackford Hill, overlooking Edinburgh, the impressive control room at Jodrell Bank beside the awesome machinery of the famous radio telescope dish. I added the backgrounds wherever I could as these placed the sitters in the context of their work and I could make shapes and compositions with them.

It was a tremendous boost that The Science Museum has acquired the complete set of portraits, and that the National Portrait Gallery and Scottish National Portrait Gallery have added astronomers to their collections. I hope that you will enjoy looking at these photographs of a selection of the leading astronomers in the UK and reading about their remarkable lives. If I have omitted to include anyone the fault is entirely mine.

Introduction
Alison Boyle

This book contains portraits of individual astronomers, but taken as a whole offers a portrait of astronomy in the UK as we celebrate International Year of Astronomy.

2009 marks four hundred years of using telescopes to explore the skies. Thomas Harriot, one of the first people to do this, could scarcely have imagined just how much things would change in the centuries after he turned his 'perspective trunke' to the Moon. Today, telescopes form just part of the dizzying range of observational, computational and theoretical techniques used by Harriot's heirs. Their thoughts touch the far reaches of our solar system, stretch beyond to faraway galaxies, and even explore extra dimensions. We still have much to learn about the cosmos, but what we do know owes a great debt to the people featured in this book.

The second half of the 20th century and start of the 21st, a period spanned by the careers of those portrayed, has seen huge advances in astronomy. Many of the men and women in these pages have pioneered these advances. Invisible-wavelength astronomy and the ability to send telescopes and probes into space have allowed astronomers to observe more celestial objects and phenomena than ever before; the rapid growth of electronic technology means that they can gather and process vast amounts of data, or test extremely complex theoretical models. As often as their work builds on existing knowledge, it can throw up new questions that take us in unexpected directions. The last few decades have also seen a boom in different forms of mass media, with the work of professional astronomers now made rapidly available to all of us. Many of the people in this book have been at the forefront of making astronomy accessible and enjoyable on a popular level.

Although the 38 portraits here reflect UK astronomy as a whole, the most enjoyable aspect of this book is the chance to get to know each individual a little. Astronomy is all too often presented as an abstract 'big science': we enjoy striking images, marvel at giant observational facilities and scratch our heads about strange-sounding theories, but we rarely encounter the thousands of individuals who have played a part in creating these. They consider extraordinary things, but they are ordinary people. Lucinda Douglas-Menzies' portraits focus on the person rather than his or her work; the text accompanying each one tells a very human story. The insightful and often self-effacing testimonies remind us that astronomers are usually made, not born. They come from a wide range of backgrounds and have taken many different paths towards their current careers. For many, an encounter with an inspirational person or book has made the difference. Hopefully, encounters with the inspirational people featured in this book will help to attract and encourage future generations of astronomers.

Curator, Astronomy and Modern Physics,
The Science Museum, London. January 2009

The Sitters

The Portraits

"I think that at some time in their lives everybody can experience that peculiar feeling of the mystery of the Universe and of the stars; of the wonder of our place in the cosmos; and of the Earth being no more – as we have discovered – than a tiny speck of dust in the vastness of space, yet a part of it. Astronomers are no different, but we are privileged to spend time exploring these questions and to contribute to our collective understanding of the Universe and the world in which we live. My earliest inspiration in this area was a book, 'The Spangled Heavens' by Lawrence Edwards. Reviewing this, even today, I am struck by the simple and yet mature way in which he introduces the Sun and the Solar System, and indeed the stars and galaxies beyond our astronomical backyard. I recall the references to what we nowadays call the 'big questions': the origins and conditions necessary for life; the ever-changing appearance of the planets; the mysteries that astronomers were then facing; and how, in space, there is always something new to be discovered – and how each new discovery presents some new problem to be solved. Mysteries attract me, and puzzles and contradictions are challenges to be resolved."

Professor Mark Bailey MBE.
Astrophysicist and Director of the Armagh Observatory. Portrait taken by the window in my studio in London 10 May 2008.

"My earliest interests in science were in chemistry, which saw the Christmas chemistry set received at age 11 bubble up into a formidable home laboratory over the next few years. Gradually, I realised that what interested me most about chemistry was the part of it that was turning into physics, and what interested me most about physics by age 15 was astrophysics: the use of physics to understand astronomical objects. I had no special interest in looking through telescopes and doing observational astronomy. Instead, I was fascinated by the fact that the same simple physics and maths that we used to understand simple aspects of gravity, thermodynamics and light, here in the mundane circumstances of the school laboratory, could successfully explain how stars and galaxies 'worked', far across the Universe, and at times far into the past when their light first began its interstellar journey towards us. I was fortunate to find that the local libraries carried a number of good elementary books on these subjects. Remarkably, one day I would come to know some of their authors quite well. And those same questions that you first ask about the Universe when you learn that scientists study it – How old is it? How big is it? What is it made of? Did it have a beginning and will it have an end? – turn out to be those that cosmologists ask still."

Professor John Barrow FRS.
Professor of Mathematical Sciences, University of Cambridge.
Taken in his office at the Centre for Mathematical Sciences,
Cambridge 24 July 2007.

"My career in astronomy nearly didn't happen! First I failed the 11+ at my school in Northern Ireland and then in the first week of secondary school the class was divided – boys to science lessons, girls to domestic science. My parents raised the roof when they heard this and the next time the science class met it consisted of three girls and all the boys. In the exam at the end of that term I came top of the class. I hope that the school learnt something from this. I certainly did. That science, especially the physics and astronomy we had been studying that term, was for me. The launch of Sputnik a few years later, in 1957, meant an increased emphasis on science in the UK and this confirmed my career choice.

A physics degree at Glasgow was followed by a PhD at Cambridge in radio astronomy. During my time there I was involved in the discovery of pulsars – work which was recognized by the award of a Nobel Prize to my supervisor. Learned bodies in the UK and the US have presented me with prizes and medals. In 2008 I became the first female President of the Institute of Physics. In demand as a speaker and broadcaster I see public engagement with science as important, and by being visible hope to encourage more women into science."

Professor Dame Jocelyn Bell Burnell DBE, FRS.
Visiting Professor of Astrophysics in the University of Oxford and current President of the Institute of Physics, the first woman to hold the title. Taken in the Department of Astrophysics, Denys Wilkinson Building, Oxford 16 November 2007.

"While visiting the Albert Einstein museum in Berne, at a time when I was wrestling with writing a particularly challenging research paper, I was very impressed with one of his statements: "Das Erfinden ist kein Werk des logischen Denkens. Wenn auch das Endprodukt an die logische Gestalt gebunden ist". It was a tremendous relief to me to realise that scientific advance was acknowledged, even by the great Einstein, not to happen via cool, calm logical steps – but implicitly with the help of intuition and imagination – even if the final product would of course be robustly defensible via logical steps. I had feared that going into science would mean a dehumanizing and detached existence, but it's been the opposite: engaging and inspiring to glean new insights into the complex and rich behaviour of the cosmos – be it quasars in the distant Universe or microquasars in our Galaxy – from the vantage point of this seemingly insignificant planet."

Professor Katherine Blundell.
Royal Society University Research Fellow and Professor of Astrophysics in the University of Oxford. Taken 1 July 2008 in the gardens of St John's College, Oxford where she is a Science Research Fellow.

"My engagement with astronomy did not arise directly from any considered choice I had made. It began merely as an opportunity for a first job, producing instrumentation for space astronomy projects at University College London, after doctoral research in atomic physics. In time I was invited to advise on instrumentation for the new Anglo-Australian Telescope nearing completion. It occurred to me (indeed in the bath) that in taking either images or spectra of distant, therefore very faint, objects in the sky a vast gain in sensitivity could be achieved by building up in a computer memory a digital record of detections of the single particles of light – photons – as they arrived one by one on a special intensifying camera, and I saw exactly how to do it efficiently with an assembly of existing devices. I was able soon to build the Image Photon Counting System or IPCS as it became known, and as my first real venture in observational astronomy took it with my expert team to the Palomar Observatory to try it out on the world's then premier telescope, the famous "200 inch". This was in the early 1970s. It was then by chance that in the corridor I came across a top astronomer on the staff, Wallace Sargent, and within seconds we agreed to collaborate. In our first run we made a major discovery about the nature of quasars and the intergalactic medium, then continued together to develop this into an important branch of cosmology. At the same time the IPCS was recognised as a revolutionary advance and was taken up by major ground-based observatories around the world and also as a main instrument for the Hubble Space telescope."

Professor Alec Boksenberg CBE, FRS.
Honorary Professor of Experimental Astronomy, University of Cambridge and Chair, United Kingdom National Commission for UNESCO. This portrait of him, holding an intensifying television camera relating to the Image Photon Counting System, was taken at the Institute of Astronomy, Cambridge 21 April 2008.

"Reading some Patrick Moore fiction when I was 8 led me to the point of no return in 1957 when: Jodrell Bank opened; uncle Joe showed me Comet Arend Roland; Sputnik and Sky at Night were launched. Around 11, I started telescope making with dad's eccentric photographer friend Eddie Cotogno, by taping old lenses onto cardboard tubes. Later I made 12 and 20 cm mirrors, building mounts with help from my engineering dad. (Though a theorist professionally, I remain keen on stargazing and linking theory to data). Backed by my History and Physics teachers, "JT" Robertson and John McIntyre, I founded Dumbarton Academy's Astronomy Society and attended evening classes by Archie Roy and Mike Ovenden in Paisley where Coats Observatory was being revitalised. The theorist awakened when I worked out for my maths teacher why there are two equal tides a day. Subsequently I studied astronomy at Glasgow, getting a faculty job at 21 and rising through the ranks there while taking taking numerous fellowships abroad. 1957 also launched me into conjuring, a further passion of mine alongside painting . Above all my enthusiasm stems from the individuals named above, and others who inspired me early on."

Professor John C. Brown FRSE.
Regius Professor of Astronomy, University of Glasgow
(Chair founded 1760) and 10th Astronomer Royal for Scotland
(Chair founded 1834). Taken in the Kelvin Building, University
of Glasgow, during a thunderstorm, 28 April 2008.

"At age four, my first view of the beauty of the stars in the summer sky during a night-time boat crossing from England to France was the earliest step toward a life time love of astronomy. Then I developed an early interest in arithmetic and in numbers (especially large ones with many powers of ten to write out and contemplate)... When I was 12 or 13 years old my grandfather gave me Sir James Jeans' popular books on astronomy. Suddenly, I saw my earlier fascination with the stars linked to my other delight in large numbers. That the nearest star is 26,000,000,000,000 miles away revived those excitements of my first school years (although falling short of my then favorite contemplation, 1 followed by 36 zeros). I decided then and there that the occupation I most wanted to engage in "when I was grown up" was to determine the distances of the stars."
Taken from 'Watcher of the Skies' by E.M. Burbidge. Annual Review of Astronomy and Astrophysics volume 32 1994

Professor Margaret Burbidge FRS.
Astrophysicist and currently Professor Emeritus of Physics at UCSD (University of California at San Diego) where she helped develop the faint object spectrograph in 1990 for the Hubble Space Telescope. Taken in the Goring Hotel while she was over here, visiting from the States 15 April 2008.

"Why did I choose astrophysics? I was the kind of child that would ask 'why does the sun always rise over there, in the same place?' I was also the kind of child that took sketchbooks on holidays. One is always constrained to make a choice, and I chose science. I think it was mathematics that did it for me, in its intrinsic elegance and in discovering the relationship between mathematics and the actual, experienced world that we live in. That, and being allowed to stay up late to watch the moon landings, at age 8; such a powerful demonstration of what can be done, what can be calculated, brought into being by our simple, honest and determined efforts to understand and quantify the universe that we find ourselves in.

So did I make the right choice, between art and science? I will happily sit for two hours on a rock making sketches and studies of a cliff face, or a coastline, because it is stunningly beautiful. I will just as happily spend long hours each day, struggling with the mathematics needed to capture some aspect of the turbulent solar wind. I take pleasure in the way colour flows on a page, and in how mathematics flows to form patterns in space. Eventually, I come up with something new, something that we have not seen before. Perhaps, there isn't such a difference between art and science after all."

Professor Sandra Chapman.
Professor of Astrophysics and Director of the Centre for Fusion, Space and Astrophysics, University of Warwick, Coventry. Taken at the Institute of Physics, Portland Place, London 23 June 2008.

"I came late to science and only studied Chemistry in my last two years at school. Following a Physics degree at University College Dublin, I attended a seminar on Space in early 1960 by Robert Boyd. Enthralled by the vision he outlined, I approached him after the talk and ended up with a Royal Society Gassiot Committee Research Studentship for Ph.D work in his group at UCL. The project involved building proportional counter X-ray spectrometers for Solar X-ray studies and flying them on Skylark rockets from Woomera. I also analysed solar X-ray data from the UK/US Ariel I spacecraft, the world's first international collaboration in space science and began preparation for an X-ray spectrometer that later flew successfully on the NASA Orbiting Solar Observatory spacecraft and established how solar flare plasma cools in the Sun's corona.

Following the Ph.D, I spent the next several decades working on both solar and cosmic X-ray astronomy. The latter was a new discipline and several outstanding discoveries emerged in the 1970s from the UK instruments on the NASA Copernicus and UK/US Ariel V missions. These were exciting and successful times. My interest in high temperature plasmas led increasingly towards high resolution X-ray spectroscopy while collaborators and friends in both USA and Japan and the manner in which space mission opportunities arose led me undertake mainly solar observations in the past 25 years.

Apart from a year at the Lockheed Palo Alto Laboratory and some leaves of absence mainly in USA and Japan, my career has been spent at UCL's Mullard Space Science Laboratory. However if the early Australian adventures are counted, much of the focus has involved the Pacific rim which has added greatly to my experience and knowledge of the world. I have been fortunate to collaborate with some outstanding people, both in the UK and elsewhere but as an experimentalist I salute the extraordinary range of space capabilities that were established at MSSL by my predecessor and which I have striven to enhance over two or more decades. Without the backing of so many excellent MSSL colleagues, life would have been much more difficult and much less fun!

Access to space is now vital for a broad range of scientific disciplines and I have been fortunate to play some role in the development of this great endeavor!"

Professor Len Culhane FRS.
Taken at the UCL Department of Space and Climate Physics,
Mullard Space Laboratory, Holmbury St Mary, Dorking, Surrey
26 June 2008.

"I was born and brought up in South Africa with my first view of Saturn being through a telescope in our front garden which my father had built himself, including grinding of the 10 inch mirror! I saw a little ball of a planet with what looked like arms next to it, which were its rings. Little did I know that years later I would be fortunate enough to have an instrument onboard the Cassini spacecraft which is orbiting around this fascinating ringed planet. After obtaining my undergraduate and PhD degrees from the University of Natal, Durban, South Africa I spent a couple of years at a Max Planck Institute in Heidelberg, Germany and then joined Imperial College London where I was able to focus on planetary sciences. I have been involved in analysing data from missions to Jupiter (the Ulysses and Galileo spacecraft), the Cassini-Huygens mission to Saturn and am presently involved in planning two new spacecraft missions back to the Jupiter and Saturn systems. I was recently awarded the Royal Society 2008 Hughes medal for the discovery made via the Cassini magnetometer instrument of an exotic and dynamic atmosphere at one of Saturn's small moons, Enceladus."

Professor Michele Dougherty.

Professor of Space Physics, Imperial College, London. Taken at The Blackett Laboratory, Prince Consort Road, London 16 May 2008. In the background is a 1/25th size model of the Cassini spacecraft which is currently in orbit studying the planet Saturn and it's moons. Michele Dougherty is responsible for the magnetometer instrument which resides on the long 11 metre boom protruding off the side of the spacecraft.

"I first became interested in astronomy as a small boy. I wanted to know what the stars were and whether one could make sense of an infinite Universe, since it seemed obvious that the Universe could not have an end. I was greatly influenced by Arthur Mee's Children's Encyclopedia, the only books in our house. There, I learnt about spiral galaxies, speculations about the origin of the solar system and of the giant Palomar Telescope. Patrick Moore's 'Sky at Night' was another inspiration – it's a shame that it is broadcast so late at night nowadays. I went on to study physics at Oxford. Towards the end of my degree, I thought about doing research in particle physics. However, in the end, I enrolled to do a Ph.D in astronomy and I am very glad that I did!"

Professor George Efstathiou FRS.
Director of the Institute of Astronomy, University of Cambridge (2004-2008) and current Director of the Kavli Institute for Cosmology, Cambridge (due to be opened summer 2009). Taken in the University Observatory Library 28 September 2007.

"When I was a school kid I never even dreamt that I would become a professional astrophysicist. I was enthralled by the sky from when my father borrowed a telescope and showed me a splendid comet. Although I was young at the time, I have very clear memories of the upper part of a sash window being lowered and the telescope being rested on it. The distant and three-dimensional nature of the sky was just magical.

Through some brilliant maths and physics teaching, I went to Manchester University to do physics and when I graduated I was all set up with a job in the electronics industry. I had never even considered research. However, that all changed when I was shown a map of the Sahara. If I did a PhD then there would be a bonus of a trip to see a very long eclipse – over seven minutes. I was hooked.

Unlike some people, I never had a plan that saw me through to where I am now. There were always choices and I could have been something very different. My approach has always been to take opportunities when they are presented and see what happens."

Professor Yvonne Elsworth.

Taken 17 April 2008 on the roof of the School of Physics and Astronomy, University of Birmingham, Edgbaston where she and her team work on the BiSON (Birmingham Solar-Oscillations Network). This consists of six globally distributed remote sites where the Sun is observed as near continuously as possible and whose instruments observe 'global' modes of oscillation that probe the deep solar interior. In the background is a 12 foot diameter dome, similar to those deployed in the network, which is used to test equipment and for the training of staff.

"One of my teachers once told me that I wanted to be an astronomer when I was age 7. I recall being intrigued that astronomers could work out the properties of stars without actually going to them. When I was a young teenager I examined the Night Sky with a small telescope and later made a reflecting telescope with which I observed and drew the lunar craters. I was on a NASA mailing list and followed much of the Space Programme, which at times was very exciting. Following a first degree in Physics at Kings College London, I did a PhD in X-ray astronomy at University College London, Mullard Space Science Laboratory. This involved designing and building detectors launched on two sounding rockets which returned data on the cosmic X-ray background. Astronomy had become, and still remains, a fascinating subject to me."

Professor Andrew Fabian OBE, FRS.
Royal Society Research Professor at the Institute of Astronomy, University of Cambridge and current President of the Royal Astronomical Society. Taken at the Institute of Astronomy 14 April 2008.

"Astronomy and cosmology confront some of the most fundamental questions in the whole of science, many of which have preoccupied humankind since the beginning of civilization: How and when did our universe begin? What is it made of? How did it acquire its current appearance? Will it end and, if so, how? It was the prospect of addressing some of these questions that attracted me to physics and astronomy. I was very lucky to have had the opportunity to become an astronomer in the last part of the twentieth century because in the past few years there has been tremendous progress in attempting to answer some of these questions. For example, recent observations have identified what the universe is made of, revealing a completely unexpected mix of not only ordinary atoms, but also exotic dark matter and a new form of energy called dark energy. Gigantic surveys of galaxies show how the universe is structured. Elegant theoretical ideas trace back the origin of galaxies to quantum processes that occurred just after the Big Bang and large supercomputer simulations relate them to the structures seen today, recreating 13 billion years of cosmic evolution. Once you have seen the sky at night and have had a glimpse of the physical laws that rule it, how can you want to be anything other than an astronomer?"

Professor Carlos Frenk FRS.
Director, Institute for Computational Cosmology, Durham University. Taken 13 November 2007 and shows in the background a map of the dark matter in the "Millennium Simulation", the biggest supercomputer simulation to date of the evolution of the Universe from the Big Bang to the present day.

"Thinking about big abstract questions, and being the first person to discover new things, is the perfect life! But I became an astronomer purely by chance. I was – and remain – interested in learning, discovering, and understanding (I'm much less good at the last one), but there are dozens of interesting subjects which allow that. I really drifted into astronomy by laziness – just continuing to do physics and maths because I was good at those – and by chance, when I realised quantum mechanics was the wrong place to make new discoveries. Archaeology, anthropology, architecture, art ... just in the A's – all appeal, and all interest. So I'm an accidental astronomer, but an interested questioner. I like discovering the new, and thinking about the fundamental. That's all, really."

Professor Gerry Gilmore.

Professor of Experimental Philosophy and Deputy Director, Institute of Astronomy, University of Cambridge. Professor Gilmore is seated on the historic Northumberland telescope (donated by the Duke of Northumberland in the 1830s). Taken at the Institute of Astronomy 14 April 2008.

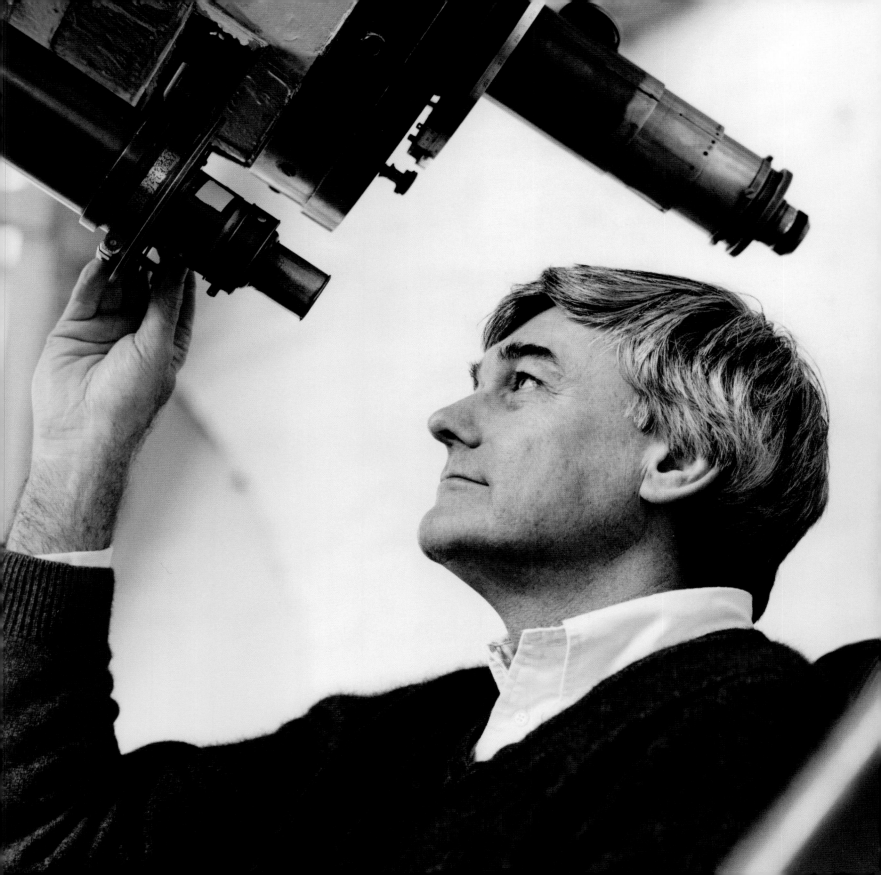

"My eureka moment came when I first saw thin-sections of Apollo Moon rocks. I was studying geology at Durham University and was used to looking at thin-sections of terrestrial rocks, in which the mineral grains are usually cracked and broken. This is caused by weathering – the effect of wind, rain and frost – and means that the grains are often pitted, with cloudy colours. The Apollo rocks were completely different – no rain on the Moon! So their minerals had deep and clear colours, with sharp and distinct grain boundaries. The grains were whole, and not broken or pitted. It was a revelation to see how different rocks from the Moon were from the terrestrial basalts with which I was familiar. I knew then that I wanted to study extraterrestrial rocks and learn more about them. I'm still learning 30 years later, and still get a buzz from seeing the colours and textures of meteorites and Moon rocks."

Professor Monica Grady.
Professor of Planetary and Space Sciences, The Open University, Walton Hall, Milton Keynes holding a meteorite that probably came from the Asteroid 4 Vesta. Taken on April 10 2008.

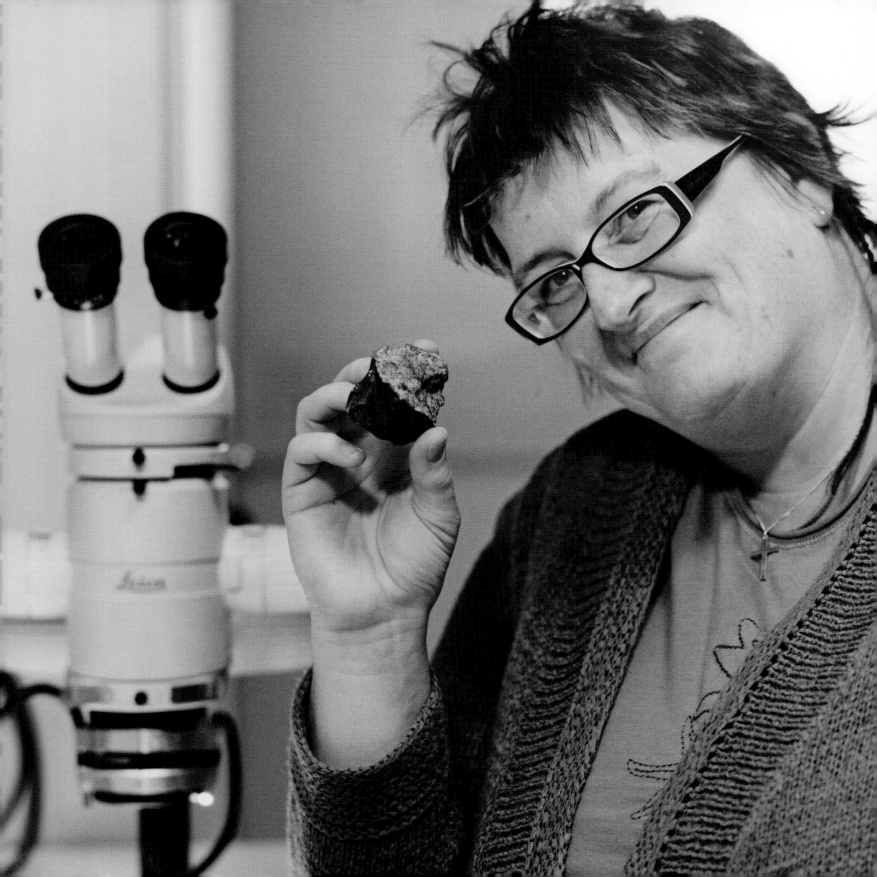

"I became an astronomer without knowing it in 1946, when I returned to Cambridge University after working on radar during the later years of World War II. Up to that time I had not been taught anything about the stars, the planets or the Universe. I started my research in the Cavendish Laboratory with Martin Ryle (who later became 12th Astronomer Royal) as a physicist with special interest in radio, based on our wartime experience. Surprisingly powerful radio waves had been detected from the Sun and from the Milky Way, beyond anything expected by conventional physics. We learned how to locate the sources of these cosmic radio emissions, some from interstellar space and some from objects that we first thought must be stars like the Sun. So I found my way into the world of classical astronomy, providing positions for several radio sources which turned out to be interesting objects such as the remains of supernova explosions, and some of the most distant galaxies then observable. More explorations followed, including an early satellite to measure radio waves from above the Earth's atmosphere. A diversion was to track the first manmade satellite. Sputnik I; this led to a proposal for a navigation system before GPS had been invented.

From this unconventional entry into astronomy, I eventually became Director of the Royal Greenwich Observatory, and afterwards the 13th Astronomer Royal. Maybe this shows the flexibility of the astronomical establishment in conferring such honours on a late entry into their world."

Sir Francis Graham-Smith FRS.
13th Astronomer Royal. Taken 14 November 2007 in the grounds of Jodrell Bank Observatory, Cheshire, where he was a Director (1980-1990).

"…in my teens I built model aeroplanes and boats….My aim was always to build working models that I could control…Since I began my PhD, this need has been met by my research into cosmology. If you understand how the universe operates, in a way you control…. When I came to the last two years of school, I wanted to specialise in mathematics and physics….My father encouraged my interest in science, and even coached me in mathematics until I got to a stage beyond his knowledge… But from the age of thirteen or fourteen I knew I wanted to do research in physics because it was the most fundamental science. This was despite the fact that physics was the most boring subject at school because it was so easy and obvious. Chemistry was much more fun because unexpected things, like explosions, kept happening. But physics and astronomy offered the hope of understanding where we came from and why we were here. I wanted to fathom the far depths of the universe. Maybe I have succeeded to a small extent, but there's still plenty I want to know." Taken from Chapter 1, Childhood, 'Black Holes and Baby Universes' and other essays

Professor Stephen Hawking CH, CBE, FRS.
Theoretical physicist and Lucasian Professor of mathematics at the University of Cambridge. Taken at the Department of Applied Maths and Theoretical Physics, Cambridge on 11 June 2008.

"My life as an astronomer, and teacher of physics, began in 1948 when I had completed my studies at Cambridge following three years war service at the Royal Aircraft Establishment, Farnborough. I had no prior ambitions to become an astronomer, but when offered a research studentship at the famous Cavendish Laboratory it was too good an opportunity to miss. So I joined Martin Ryle's small group which was just beginning to investigate radio waves from the first "radio stars". I had worked with Ryle before in 1944 when seconded to the Telecommunications Research Establishment (TRE), Malvern. Ryle was then involved with radar jamming devices required by R.A.F. Bomber Command. My job was to visit the station where the equipment was first installed and provide instruction.

For my Ph.D. project I studied the scintillation of radio sources caused by clouds in the ionosphere. This gave valuable information at heights far above those probed by other methods. In the 1960s I developed similar methods to measure the speed of the solar wind, which blasts outwards through interplanetary space at around one million miles per hour. We had found that the energetic radio galaxies called quasars, due to their tiny angular size, showed strong scintillation caused by the solar wind, and in 1967 I began a sky survey, based on this effect, to reveal more quasars. These observations required a special radio telescope, covering an area of 4.5 acres and containing 2048 dipole antennas, which I had designed in 1965.

By an amazing stroke of good fortune it turned out that the requirements for the quasar search were ideally suited to reveal the first pulsars. My graduate student Jocelyn Bell-Burnell, who was responsible for analyzing the data, found a source showing scintillation that was unusually strong, and our more detailed observations revealed the clock-like pulsed radiation that is characteristic of neutron stars. This opened an exciting new chapter of astronomy which remains highly active today."

Professor Antony Hewish FRS.
Nobel Laureate in Physics (1974): radio astronomer.
Taken 6 July 2007 in the control room of the One Mile Telescope, designed by the late Martin Ryle, at the Mullard Radio Astronomy Observatory at Lord's Bridge near Cambridge. The name, One Mile, reflects the fact that the three dishes are equivalent to a single giant reflector one mile in diameter.

"A remark by Sir Arthur Eddington "An observation cannot be believed until it has been proven by theory", inspired me to apply my mathematical studies at University to the challenging problems of the physical world. My first opportunity came in Oxford in the mid 1960s when working with Sir John Houghton and Professor Ian Grant in developing models to obtain meteorological information from the early weather satellites. It was already becoming clear to us, that the Earth's climate was changing significantly.

I became curious about our planetary neighbours, extending my studies to the other atmospheres in the solar system, initially Venus, Mars, Jupiter and Saturn. I realised this was an opportunity to increase our basic knowledge of these bodies and from the very limited telescopic observations I had a unique opportunity to conduct comparative studies of the Earth and planets, whose insights have helped identify the issues and challenges of global warming.

After joining the Jet Propulsion Laboratory, in Pasadena, California in the early 1970s, I was able to apply my work to NASA's rapidly developing planetary projects. I became a Principal Investigator on the NASA Voyager mission to the outer planets, which has been the highlight of my life and lasted over twenty years."

Professor Garry Hunt.

Professor of Atmospheric Physics, photographed at home in West Wimbledon 21 June 2007. On his laptop can be seen a view of Saturn where Professor Hunt is studying the varied cloud patterns and weather systems of this giant planet.

"Although I studied physics and mathematics at the school I attended, I did not find A-level physics particularly exciting and was unsure what to study at University. At school I came across a popular book on astronomy by Eddington and rapidly decided that physics in the context of astronomy was more interesting than physics alone. At the time, the only astronomy course in England was at University College London, so I arranged a visit, applied and to my delight was accepted. Because there were only a few students on the course, the environment at Mill Hill was dominated by research, and the majority of students went on to research degrees and careers in astronomy. For my PhD project at UCL I worked on modelling the solar transition region and corona from the early uv and X-ray spectra available. This included solving the problem of the origin of a group of strong solar emission lines between 170 and about 240 A (3p - 3d transitions in Fe IX to Fe XIV). My supervisor (C.W. Alien), attending an open day at Culham Laboratory, noticed that a similar set of lines were unidentified in the spectrum of the early fusion device ZETA. This led to collaboration and my spending 10 years at Culham Laboratory. In 1976 I moved to the University of Oxford, where I have remained since, and have continued research on problems relating to the hot outer atmospheres of cool stars."

Professor Dame Carole Jordan DBE, FRS.
First ever female president of the Royal Astronomical Society (1994- 1996), Head of Theoretical Physics, the Rudolf Peierls Centre for Theoretical Physics, Oxford (2005-2008). Taken in the nearby gardens of The University Parks 16 November 2007.

"My involvement in astronomy began in earnest when I joined the radio astronomy group in the Cavendish Laboratory as a research student in 1963. I was inspired by the enthusiasm of Martin Ryle about the remarkable opportunities for innovative science in the relatively new field of radio astronomy. Since then, my interests have expanded to cover many areas of high energy astrophysics and astrophysical cosmology. I was appointed the ninth Astronomer Royal of Scotland in 1980, as well as the Regius Professor of Astronomy, University of Edinburgh, and the director of the Royal Observatory, Edinburgh. I was head of the Cavendish Laboratory from 1997 to 2005. I have served on and chaired many international committees, boards and panels, working with both NASA and the European Space Agency. My main research interests are in high energy astrophysics and astrophysical cosmology."

Professor Malcolm Longair CBE, FRS, FRSE.
Jacksonian Professor of Natural Philosophy and Director of Development, Cavendish Laboratory, Cambridge. Taken in the museum at the Cavendish Laboratory 28 September 2007. The desk (right) belonged to James Clerk Maxwell, (1831-1879), the first Cavendish Professor of Experimental Physics. The apparatus (left) is Maxwell's experiment to determine the dependence of the viscosity of gases upon pressure and temperature, a key test of the kinetic theory of gases.

"In 1939 I was a physicist working on cosmic rays with Blackett in the University of Manchester. In August of the year when war was imminent I was sent to the radar research establishment at Bawdsey Manor and when Prime Minister Chamberlain announced that we were at war with Germany I was in the operation room of the radar defense station at Staxton Wold, Yorkshire. The radar echoes on the cathode ray tube were not from enemy bombers but were said to be from the "ionosphere". I thought they might be echoes from the highest energy cosmic ray particles entering the earth's atmosphere and after six years of war I borrowed some army radar equipment to investigate these phenomena. In December 1945 that was the beginning of the astronomical part of my career at Jodrell Bank. It was these investigations that led to the construction of the radio telescope at Jodrell Bank. In that way my career as an astronomer began by chance."

Sir Bernard Lovell OBE, FRS.
Physicist and radio Astronomer. Taken 14 November 2007 in the control room at Jodrell Bank Observatory, Cheshire, with the Lovell Telescope in the background, constructed with University funding. When it was built in the mid 1950's the telescope was the largest steerable dish radio telescope in the world at 76.2m (250ft) in diameter; even now it is the third largest, after the Green Bank and Effelsberg telescopes and forms part of the MERLIN and European VLB1 Network arrays of radio telescopes.

"My father was interested in Science and inherited from his grandfather a fine 3 1/2 inch brass telescope. This gave me a keen amateur interest in Astronomy. I was always interested in how things worked and relativity, quantum theory and statistical mechanics were all of philosophical interest. I aimed to go into particle physics but when I was ready to start all the interesting lecturers in that subject were away so I turned my hobby into my profession. By luck this was a good choice."

Professor Donald Lynden-Bell CBE, FRS.
Emeritus Professor of Astrophysics in the University of Cambridge. Best known for his theories that galaxies contain massive black holes at their centre and that such black holes are the principal source of energy in quasars. Taken in his study at the Institute of Astronomy, Cambridge 21 April 2008.

"My early interest in astronomy, stimulated e.g. by the 'Children's Encyclopaedia', continued into my teens, when the first 'Pelican' paperbacks I bought included 'The Stars in their Courses' by Sir James Jeans and 'The Expanding Universe' by Sir Arthur Eddington; but as yet, I had not thought of making a career in astronomy. In my third year at Cambridge I opted for the courses in Theoretical Physics offered for Part III of the Mathematics Tripos. These included Hermann Bondi on General Relativity and Cosmology, Dirac's monumental treatise 'Quantum Mechanics', Nick Kemmer on Nuclear Physics and Quantum Field Theory, and Fred Hoyle on Statistical Thermodynamics. I toyed with the idea of working for a Ph.D. in Field Theory, but, these were the days before Feynman et al.: in Nick's words, 'We are waiting for a new Heisenberg' – enough to deter all but the most intrepid. Fred Hoyle included in his wide-ranging course the theory of degenerate matter applied to white dwarf stars, and an account of his first paper on cosmical nucleogenesis. It was clear that astrophysics offered scope to a young newcomer, with interesting problems that were both important and solvable. Fred agreed to take me on as his Ph.D. student, and later I was a post-doc, with Tom Cowling in Leeds and with Lyman Spitzer and Martin Schwarzschild in Princeton. I never looked back."

Professor Leon Mestel FRS.

Emeritus Professor of Astronomy in the University of Sussex, whose astrophysical research interests include stellar structure and evolution, star formation and cosmical magnetism, especially pulsar electrodynamics. Taken at the University of Sussex, Brighton 4 April 2008.

"I became interested in astronomy at the age of seven. My mother had a casual interest in it, and had a few books about it; by chance I picked one up – 'The Story of the Solar System' by G.F.Chambers (published in 1898) and was fascinated.

I did some more reading (I was able to cope with adult books), obtained a star map and learned my way around the night sky. Binoculars next; I joined an astronomical society, and went on from there. At the age of 11 I was lucky enough to be made a member of the BAA.

The idea was Eton and then Cambridge, to take my degree, but it didn't work out like that. Until I was 15 I wasn't fit enough for school; at 16 I could get around and had my Cambridge place, but Herr Hitler intervened, and I joined the RAF to fly. So I remain an amateur astronomer, which suits me very well."

Sir Patrick Moore CBE, FRS.

Amateur astronomer who has attained prominent status in astronomy as a writer, researcher, radio commentator and television presenter of the subject and who is credited as having done more than any other to raise the profile of astronomy among the British general public. He is the presenter of the longest running television series (with the same original presenter), The Sky at Night, on the BBC which has just celebrated its fiftieth anniversary. Taken at his home in Selsey, West Sussex 5 June 2007. On the desk is an orrery presented by the British Astronomical Association to celebrate 70 years of membership.

"I was inspired by an early love of mathematics and physics and read physics and astronomy at Glasgow University but my passion for astrophysics was fired when conducting my Ph.D. at Jodrell Bank Observatory where I used the world's leading radio telescopes to study the physics of "cosmic indigestion" exhibited by supermassive black holes at the centres of active galaxies.

Later on, as a Royal Society University Research Fellow at Liverpool John Moores University my research broadened to include the study of Gamma Ray Bursts and, subsequently, I established the Liverpool Gamma Ray Burst Group.

Gamma Ray Bursts are the most instantaneously powerful explosions in the Universe, created when massive stars in the distant Universe reach the end of their lives, their cores collapsing to form black holes and, in the process, releasing enormous amounts of energy – more than the Sun emits in its lifetime – in a few seconds. The rapidly fading light of a GRB is the observational key used by astronomers to unlock the fundamental physics that drive these prodigious explosions. Exploiting the innovative technology of the robotic Liverpool Telescope to study the birth of these black holes in real time, my team are on-call day and night, armed with mobile phone and computer, ready to respond instantly to alerts from satellites that beam messages to Earth, announcing the occurrence of a new GRB and triggering a flurry of activity by astronomers across the globe."

Professor Carole Mundell.

Professor of Extragalactic Astronomy, Astrophysics Research Institute, Liverpool John Moores University. Taken 29 April 2008 at her home in Buckley, Flintshire, with her computer and mobile phone at the ready to respond to any alerts from satellites of the occurrence of a new GRB.

"I became a professional astronomer by accident. Of course I looked at the sky and wondered about it when I was young, but no more than anyone else. I certainly had a broad interest in science from early on, driven by a mixture of the factual writings of Isaac Asimov and a variety of science fiction authors. At school, however, I found physics boring, and settled on studying chemistry. Fortunately, I went to Cambridge, where chemists have to do a year of physics. By the end of this, I'd realised that all the things I liked most in Chemistry (such as atomic structure) were really physics, so I switched. At the end of a physics degree, I looked for jobs; these largely seemed to consist of making radars for the military, so I decided to keep my options open by doing a PhD. I should say that a significant factor in this decision was music: I am an amateur clarinettist, and the standard of Cambridge student music is extremely high, so staying in that world was pretty attractive. In fact, I did very little work in the first year of my PhD, but I did do a lot of concerts. I settled on astronomy research almost by elimination, since most of the other options on offer seemed too narrowly specialised. So it was really only after a slow start that I came to realise how much I enjoyed working in astronomy. I started as a postdoc in 1981, just as there was an explosion of new ideas in cosmology. Being swept along by that process, and being able to contribute to it, was tremendously exciting. I feel extremely fortunate to have ended up where I am now, and I wouldn't change a thing."

Professor John Peacock FRS, FRSE.

Professor of Cosmology and Head of the Institute of Astronomy, University of Edinburgh, photographed at the Royal Observatory, Blackford Hill, Edinburgh 28 April 2008. In the background is a picture showing the distribution of galaxies as seen in the 2dF (Two-degree Field) Galaxy Redshift Survey which was a UK-Australia collaboration of over 30 astronomers, of which he was the UK Chairman.

"My early fascination for astronomy came from my father (Galton Professor of Human Genetics, UCL), who knew many constellations and stars. In 1927 (four years before I was bom), he and my mother witnessed a total solar eclipse, using an old brass telescope he owned. When I was a teenager, he showed me Saturn, and telescopic verification that its beautiful rings were real had a great emotional effect on me! In cosmology, the first big influence was the thrilling series of radio talks, given in the 1950s by Fred Hoyle, which led up to a wonderful description of the new Steady-State model that he, Gold, and Bondi, had recently developed. Though finding this fascinating and emotionally gratifying, I was disturbed by Hoyle's statement concerning distant galaxies disappearing from view when their recession speeds reached that of light. By drawing space-time diagrams, I convinced myself it was not quite accurate. Informing the young cosmologist Dennis Sciama (a friend of my brother Oliver's), we formed a lasting relationship whereby I learnt much cosmology and physical science. The elegant Steady-State model, though subsequently discredited by observation, provided an important input into my later thinking."

Sir Roger Penrose OM, FRS.

Mathematical physicist and Emeritus Rouse Ball Professor of Mathematics in the University of Oxford. He is renowned for his work in mathematical physics, in particular his contributions to general relativity and cosmology and is also a recreational mathematician and philosopher. Taken at the Mathematical Institute, Oxford 4 March 2008.

"My father was a physicist with a research interest in cosmic rays, using cloud chambers in the lab which led to an interest in cloud formation in the atmosphere. His enthusiasm for sky-watching was infectious. We lived in Edinburgh where aurorae were a (fairly) common sight and I also spent many holidays in mid-Wales with the most fantastic dark skies. I longed to learn more. I studied physics at Edinburgh University and spent one vacation at the Royal Greenwich Observatory (RGO) where I met my future husband. It was the start of a life-long connection with the RGO until it closed in 1998. My own interest is in 'nearby' stars (that is, stars within about 300 light years of the Sun in our corner of the Milky Way Galaxy). I am most proud to have been involved in the European Space Agency (ESA) Hipparcos project. We obtained the distances, motions and brightnesses of 118,000 Milky Way stars, producing a fantastic database for further work. Astronomers' knowledge of the most distant objects in the universe is based on such work. I have always been interested in promoting astronomy in the UK and have had a long involvement with the Royal Astronomical Society, including periods as Vice President and Astronomy Secretary. I was also President of the Society for Popular Astronomy for two years, which helps some of the thousands of amateur astronomers in the country to develop their interest in the subject."

Dr Margaret Penston MBE.
Taken at the Institute of Astronomy, University of Cambridge on 24 July 2007.

"I was actually trained as a physicist and a career in astronomy began by chance when I was offered a scholarship to read for a PhD at UCL, working under Robert Boyd in the new Rocket Research Group. My thesis project was to develop a method of measuring the solar X-ray spectrum from space, supporting a broader study of the effects of solar radiation on the Earth's upper atmosphere.

From UCL I moved to Leicester in 1960 to help start a new research group specialising in X-ray Astronomy. Initially the Sun was our only target and we were able to get X-ray spectra of increasing quality and the first direct images of the solar corona using the Skylark rocket and the Ariel 1 satellite.

However, following the discovery by US astronomers of the first cosmic X-ray source in 1962, the field widened dramatically and we were well-placed to make an early entry to a discipline now recognised as one of the main observational pillars of modern-day astrophysics.

Looking back over half a century I have been privileged to work alongside many talented scientists and engineers from across the world. Many have become lasting friends. A career in astronomy and space science has offered particular challenges – and pleasures – for someone, like me, working from the UK.

Long may such opportunities continue!"

Professor Ken Pounds CBE, FRS.

Emeritus Professor of Space Physics, University of Leicester photographed at the Department of Physics and Astronomy on 17 April 2008. The hardware at the front of the background is part of the mechanical assembly for the Wide Field Camera, a UK telescope which was flown on the German satellite ROSAT in 1990. Over the following year the WFC, for which Professor Pounds was the Project Scientist, carried out the first all-sky survey in the Extreme Ultra Violet. Further back are two X-ray mirrors which were developed as part of the ESA X-ray Observatory, XMM- Newton, another mission which he was involved with, launched in 2000 and still operational. The mirrors are formed of thin shells of nickel, coated with gold to improve their reflectivity.

"In my schooldays I had no idea that I would ever become an astronomer. I did, however, specialise in science – but mainly because I was bad at languages. But I turned out to be good at maths and took my degree in that subject. However, I realised I wasn't cut out to be a mathematician. I wanted to do something that involved a more 'synoptic' approach – trying to make sense of a broad range of data rather than pursuing a deductive argument. I thought of doing economics (which I still think I could have enjoyed), but got a research studentship at Cambridge in the Applied Maths department. I then had two big pieces of good luck. First, my assigned supervisor was Dennis Sciama, an immensely enthusiastic and stimulating man, who within a year convinced me that astrophysics was a good choice of subject. The second piece of luck was that in the 1960s, when I started research, the field was opening up – the first strong evidence for the big bang, for black holes, and so forth. When a subject is new, we are all 'beginners': young scientists aren't handicapped compared to those who are older and more experienced. So it was easy to make a mark quickly. But what has been wonderful about astronomy is that the pace of novel developments and discovery hasn't slowed up: there has been at least as much excitement within the last five years as in any previous time. And astronomy has another important 'plus' – it's not just of interest to a few fellow-specialists, but one had a chance to engage with a wide and interested public, in this country and worldwide."

Lord Martin Rees OM, PRS.

15th Astronomer Royal, Professor of Cosmology and Astrophysics at the Institute of Astronomy, University of Cambridge and also Master of Trinity College, Cambridge. Taken 6 September 2007 at the Royal Society, London.

"What drew me into astronomy was Fred Hoyle's wonderful book 'Frontiers of Astronomy', which I read as a student and which showed what a wonderful diversity of scientific problems there are ranging from the solar system to the universe. I trained as a mathematician and I've generally worked at the interface between theoretical models and observations. This drew me into major observational programmes, both with ground-based telescopes, which tend to be located on beautiful remote mountain tops, and with space missions, which have an intense excitement of their own.

My main work has been in cosmology, where I was particularly interested in cosmological parameters and the distance scale (I wrote a textbook on Cosmology and a monograph on The Cosmological Distance Ladder), and in the emerging fields of far infrared and submillimetre astronomy. I had the experience several times of looking at the universe for the first time in a new waveband, so I know I have lived at a special time for astronomy. I felt like Keats' 'watcher in the skies' ... 'Silent on a peak in Darien'."

Professor Michael Rowan-Robinson

Head of the Astrophysics Group at Imperial College, London (1993-2007) and President of the Royal Astronomical Society (2006-2008) photographed at the Department of Physics, Imperial College, London 28 June 2007. Behind him are optical images of prominent infrared galaxies.

"In the year 1963, Cambridge was in its best spring colours. The thoughts of an undergraduate were diverted. I should have been studying for my mathematics courses. Instead I stumbled by hazard into the back of a classroom where cosmologist Dennis Sciama was lecturing to a spell-bound audience. The topic was Mach's principle. I need not have stayed there as it was an advanced course and not part of my required curriculum. But I could not leave. The lecture, and its sequel, was of a stunning simplicity, opened up a vast new perspective, and converted me to cosmology for life. This is the moment when I decided to become a cosmologist. Of course I first had to obtain a mathematics degree, and then move to Harvard to qualify as an astronomer, before obtaining carte blanche to pursue the origins of the galaxies to the beginning of time, and even beyond. I am currently Saviiian Professor of Astronomy at the University of Oxford, where I also direct the Beecroft Institute for Particle Astrophysics and Cosmology."

Professor Joseph Silk FRS.

Professor Silk came to the UK in 1999 to take up the position of Chairman of the Oxford Astrophysics Department following a thirty year career at the University of California, Berkeley. He is an expert on galaxy formation, structure and clustering. Taken 16 November 2007 in the Department of Physics, Oxford.

"I grew up in South Africa, where the skies were clear and the Milky Way, the Magellanic Clouds and a host of stars were visible at night. This display excited my early interest in astronomy and I was encouraged by reading James Jeans's books. At Cambridge, I studied physics and chose geophysics as a research topic. I started thinking about the origin of the Earth's magnetic field – but, after a while, I decided that the Sun's magnetic field was more interesting. As a postdoc, I learnt how to tackle what was then large-scale computation. By good fortune, I was appointed to a Lectureship in the Department of Applied Mathematics and Theoretical Physics at Cambridge, which proved an ideal base for applying numerical experiments to problems in astrophysical fluid dynamics, and relating the results to analytic treatments. I have worked with colleagues and students both on individual magnetic features (such as sunspots and starspots) and on the origins of global magnetic activity, both on the Sun (where fine structure can actually be observed, from the ground or from space) and in other stars that are much more active."

Professor Nigel Weiss FRS.

Professor Weiss has a particular interest in the field of astrophysical and geophysical fluid dynamics. Taken at the Department of Applied Mathematics and Theoretical Physics, Centre for Mathematical Sciences, Cambridge 11 June 2008. The poster image in the background is of the Sun's surface, showing a sunspot and smaller magnetic features, as well as photospheric granulation, and was obtained at La Palma in the Canary Islands, with the 1m Swedish Solar Telescope of the Royal Swedish Academy of Sciences.

"As a precocious but rather solitary child from a rural one-teacher school, I spent many hours roaming the woods and fields of South-East Cornwall. I was fascinated by the natural world, by the beauty I saw both in individual plants and animals and in the way their diversity is embedded in a harmonious whole. Later, boarding school taught me to love mathematics and music, and to wonder at the connections between these human constructs and the ordering of the physical world. After a mathematics degree, I wanted to dig deeper into these issues, so I looked around Cambridge for research opportunities. Scientists at the Institute of Astronomy studied stars and galaxies from offices which faced trees, lawns, and a host of golden daffodils (planted by IoA's visionary founder, Fred Hoyle). The resonance with my childhood was all the convincing I needed. My doctoral work showed me that mathematics can be used to recreate cosmic evolution in a supercomputer, compressing billions of years into a few hours or days, so that the birth and death of "virtual" galaxies can be followed quasi-experimentally. Here was a way to confront human mental constructs with cosmic complexity on the grandest scale, and to find out if mathematics also describes the heavens. My path was set."

Professor Simon White FRS.

Managing Director, Max Planck Institute for Astrophysics, Garching, Germany photographed in my studio in London 26 July 2008. The poster in the background shows images of the so-called Millenium Simulation, a supercomputer simulation of the growth of large-scale structure in the Universe. The bright regions are places where there is predicted to be lots of material and the dark regions are places where there is very little. The calculations followed how gravity makes structure grow as the Universe ages.

"Although always interested in astronomy it was not until I was halfway through my academic career as a physicist that I started to take a professional interest in the subject. This arose because, by way of my research subject – the cosmic radiation – I started to wonder 'where they came from' as well as following my long term work on 'what they do when they arrive'.

The transition was an ideal one and I was able to view astronomical problems from a different perspective. Such was my enthusiasm that I managed to convince many Durham University colleagues to transfer their interests into astronomy. We now have one of the best Departments in the World working on virtually all areas of the subject, from theoretical cosmology to astronomical instrumentation."

Sir Arnold Wolfendale FRS.

Emeritus Professor of Physics, Durham University and Former Astronomer Royal (14th Astronomer Royal, 1991-1995). Taken on 20 September 2007 at the Royal Society, London. Sir Arnold is holding a replica of a reflecting telescope designed and developed by Sir Isaac Newton (1642-1727). The earliest telescopes, such as those used by Galileo, consisted of glass lenses, but Newton's telescope used mirrors instead to bring light to focus.

"My Eureka moment came in July 1961. While a schoolboy in North London, I found myself standing a few feet away from Yuri Gagarin when, on his world tour, after becoming the first human in space, he visited Karl Marx's grave in Highgate Cemetery. That was the start of my fascination with space. Growing up through the 1960s, I watched most of the space missions on a grainy black & white TV set and was caught up by the drama of it all. After doing a Physics degree in Cambridge, I found that I could do a PhD at University College London – they would let me design, build and launch large rockets from Woomera in the Australian desert carrying x-ray telescopes. This gave us a few minutes of observation above the Earth's atmosphere before the payload parachuted (or sometimes crashed) back to the Earth. I was hooked – I've spent the rest of my career doing space astronomy or exploration. I've contributed to the Hubble Space Telescope, the Giotto mission to Halley's Comet and the Cassini/Huygens mission to the Saturnian system, and am now working on Europe's next mission to Mars."

Professor John Zarnecki.

Professor Zarnecki is Professor of Space Science at The Open University, Milton Keynes, and has taken part in several high profile space probe missions and is an expert on space debris, space dust and impacts. Taken 10 April 2008 in the Planetary and Space Sciences Research Institute at The Open University. Behind him is a ¼ scale model of the European Space Agency's Huygens probe, launched in 1997, during which he was Principal Investigator for one of the six scientific instruments carried onboard. After a 3.5 billion km journey, it parachuted to the surface of Titan, Saturn's largest moon, in January 2005.

Afterword

We are in excellent company. No more need be said.

Yet nothing could be less clubby, more basic and universal, than to stare at a star; and it is human to acknowledge such things. Lucinda's forbear, the Rev. Lewis Evans (1755-1827), a distinguished amateur astronomer, used to sit through the night at the bottom of a dry well, where starlight alone could reach him. Intentness of gaze is a capacity we are all born with fully-charged; and the quality of contemplation is not to be measured by timing an exposure.

Portraiture and Astronomy both aim to capture something, but depend equally upon a willingness to be absorbed by wonder. Although the achievement of excellence depends on being able to make a memorable cut, part of the interest of this collection derives from its reflection of the process of recognition, which these sitters have shared with this artist.

'Le silence éternel des espaces infinis, m'effraie'[1] – in French, because something is lost in translation, and because it is hard to imagine an English astronomer saying such a thing. But that is not to deny the place of awe, if not of fright, in raising one's eyes to a heavenly body. Photography has shown how to look at everything as an image, hence without fear or favour. And hazard has come with it: now the whole idea of a subject can go out of focus: for although an image is a thing (icons notwithstanding), a subject is not.

The particular value which characterises 'Portraits of Astronomers' is more attitude than attribute, a special way of relating to a peculiar kind of object, the stellar unknown other. Here, the chill of fear and the heat of desire may be outside the frame, but what is being offered is not less personal for that. For sure, astronomers achieve greatness by being objective, by capturing the secrets of the stars and bringing down to earth's surface the laws that govern the universe: but it is possible also to discern through the pages of this guide something of the unselfconscious love which has nourished their long, strange familiarisation with the infinities outside our own solar system.

[1] 'The eternal silence of these infinite spaces frightens me'. Pascal, Pensées, 206.

Exhibition venues and dates

Launch 12 December 2008
Royal Astronomical Society
Burlington House
Piccadilly
London W1J 0BQ

2-21 March 2009
Cambridge Science Festival
The Michaelhouse Centre
St Michael's Church
Trinity Street
Cambridge CB2 15U
01223 309167
Open Mon-Sat 8am-5pm

21 March 2009 for 6 months
Ten portraits in room 38a
The National Portrait Gallery
2 St Martin's Place
London WC2H 0HE
020 7306 0055
Open daily 10am-6pm
Late night open
Thurs and Fri until 9pm

6-18 April 2009
Edinburgh Science Festival
City Arts Centre
1-3 Market Street
Edinburgh EH1 1DE
0131 529 3993
Open Mon-Sat 10am-5pm
Sun 12pm-5pm

7-28 September 2009
The Open University
Walton Hall
Milton Keynes MK7 6AA
01908 274066

1-31 October 2009
The Royal Society
6-9 Carlton House Terrace
London SW1Y 5AG
020 7451 2606
By appointment

20 November 2009-
9 January 2010
Armagh County Museum
The Mall East
Armagh
BT61 9BE
028 3752 3070
Open Mon-Fri 10am-5pm
Sat 10am-1pm 2pm-5pm

Date to be confirmed
The Ogden Centre
Durham University
South Road
Durham
DH1 3LE

Astronomy at the Science Museum

The astronomy collections of the Science Museum and its Library span from the 10th to the 21st centuries and contain over 2000 objects, books, prints and photographs. Highlights of the collection include Copernicus' 'On the Revolutions of the Heavenly Spheres', the six-foot mirror from the great Rosse Telescope, and the DRIFT-1 dark matter detector.

'Cosmos & Culture', a temporary exhibition marking International Year of Astronomy, opens at the Museum on 23 July 2009. The exhibition traces 400 years of telescope technologies, explores our changing perceptions of our place in the Universe, and examines the role astronomy has played in shaping our everyday world.

The Science Museum
Exhibition Road
London SW7 2DD
www.sciencemuseum.org.uk

Copies of this book can be ordered from
www.douglas-menzies.com

lucinda@douglas-menzies.com
Tel 07932 626284
Fax 020 7247 5538